Estimating the Benefits of the Air Force Purchasing and Supply Chain Management Initiative

Jeremy Arkes, Mary E. Chenoweth

Prepared for the United States Air Force
Approved for public release; distribution unlimited

The research described in this report was sponsored by the United States Air Force under Contracts F49642-01-C-0003 and FA7014-06-C-0001. Further information may be obtained from the Strategic Planning Division, Directorate of Plans, Hq USAF.

Library of Congress Cataloging-in-Publication Data

Arkes, Jeremy.
Estimating the benefits of the Air Force purchasing and supply chain management initiative / Jeremy Arkes, Mary E. Chenoweth.
p. cm.
Includes bibliographical references.
ISBN 978-0-8330-4188-3 (pbk. : alk. paper)
1. United States. Air Force. 2. Inventory control. 3. Business logistics.
I. Chenoweth, Mary E. II. Title.

UG633.2.A75 2007
358.4'16210973—dc22

2007041756

Published 2008 by the RAND Corporation
1776 Main Street, P.O. Box 2138, Santa Monica, CA 90407-2138
1200 South Hayes Street, Arlington, VA 22202-5050
4570 Fifth Avenue, Suite 600, Pittsburgh, PA 15213-2665
RAND URL: http://www.rand.org
To order RAND documents or to obtain additional information, contact
Distribution Services: Telephone: (310) 451-7002;
Fax: (310) 451-6915; Email: order@rand.org

Preface

One element of the U.S. Air Force's quest to transform its logistics processes to better support the warfighter is the implementation of purchasing and supply-chain management (PSCM). The goals are to reduce supply-chain operating costs and improve warfighter readiness through customer-centric processes that link demand and supply planning, purchasing, inventory management, and suppliers. But how can the success or failure of such efforts be gauged when "all other things" are not being "held constant"? Other factors have been changing at the same time that PSCM has been implemented, and some of these may also have affected outcomes linked to these goals.

RAND Project AIR FORCE has addressed this challenge by developing an econometric model to account for such "co-occurring" factors. This monograph describes the resulting method and illustrates it with a National Item Identification Number (NIIN)–level analysis of quarterly mission capable (MICAP) incidents in the context of Air Force Materiel Command's (AFMC's) own PSCM initiative.

This monograph should be of interest to anyone concerned about Department of Defense PSCM-related spending analyses, particularly for Air Logistics Centers and the Defense Logistics Agency. We hope this research will also assist AFMC's Commodity Councils, which have commandwide responsibilities for developing purchasing supply strategies.

This work was sponsored by the U.S. Air Force Deputy Chief of Staff for Logistics, Installations, and Mission Support, Directorate of Transformation (AF/A4I) and the Deputy Assistant Secretary for Contracting, Office of the Assistant Secretary of the Air Force for

Acquisition (SAF/AQC). The research was conducted in the Resource Management Program of RAND Project AIR FORCE and was part of a research project that began in fiscal year 2005, "Performance-Based Supplier Relationships and Purchasing and Supply Chain Management Baseline Measurement."

Similar RAND work for the Air Force has been documented in the following:

- *Developing Tailored Supply Strategies*, by Nancy Y. Moore, Clifford A. Grammich, and Robert Bickel (MG-572-AF, 2007)
- *Implementing Purchasing and Supply Chain Management: Best Practices in Market Research*, by Nancy Nicosia and Nancy Y. Moore (MG-473-AF, 2006)
- *F100 Engine Purchasing and Supply Chain Management Demonstration: Findings from Air Force Spend Analyses*, by Mary E. Chenoweth and Clifford Grammich (MG-424-AF, 2006)
- *An Assessment of Air Force Data on Contract Expenditures*, by Lloyd Dixon, Chad Shirley, Laura H. Baldwin, John A. Ausink, and Nancy F. Campbell (MG-274-AF, 2005)
- *Using a Spend Analysis to Help Identify Prospective Air Force Purchasing and Supply Management Initiatives: Summary of Selected Findings*, by Nancy Y. Moore, Cynthia Cook, Clifford Grammich, and Charles Lindenblatt (DB-434-AF, 2004)
- *Implementing Performance-Based Services Acquisition (PBSA): Perspectives from an Air Logistics Center and a Product Center*, by John Ausink, Laura H. Baldwin, Sarah Hunter, and Chad Shirley (DB-388-AF, 2002)
- *Implementing Best Purchasing and Supply Management Practices: Lessons from Innovative Commercial Firms*, by Nancy Y. Moore, Laura H. Baldwin, Frank Camm, and Cynthia R. Cook (DB-334-AF, 2002)
- *Federal Contract Bundling: A Framework for Making and Justifying Decisions for Purchased Services*, by Laura H. Baldwin, Frank Camm, and Nancy Y. Moore (MR-1224-AF, 2001)

- *Performance-Based Contracting in the Air Force: A Report on Experiences in the Field*, by John Ausink, Frank Camm, and Charles Cannon (DB-342-AF, 2001)
- *Strategic Sourcing: Measuring and Managing Performance*, by Laura H. Baldwin, Frank Camm, and Nancy Y. Moore (DB-287-AF, 2000).

RAND Project AIR FORCE

RAND Project AIR FORCE (PAF), a division of the RAND Corporation, is the U.S. Air Force's federally funded research and development center for studies and analyses. PAF provides the Air Force with independent analyses of policy alternatives affecting the development, employment, combat readiness, and support of current and future aerospace forces. Research is conducted in four programs: Aerospace Force Development; Manpower, Personnel, and Training; Resource Management; and Strategy and Doctrine.

Additional information about PAF is available on our Web site: http://www.rand.org/paf/

Contents

Figures and Tables

Figures

Tables

Summary

Whenever an organization implements a new or revised process, it needs to know whether that process is achieving the expected outcomes. But because other processes may be changing at the same time, it can be difficult to determine how much of the overall outcome is attributable to that process or to some combination of events. The Air Force faces just such a problem as it strives to transform its logistics processes to better support the warfighter through its Expeditionary Logistics for the 21st Century (eLog21) program.

An initiative implemented at the Air Force Materiel Command, called Purchasing and Supply Chain Management (PSCM), is one of the tools being use to reach this objective. The goals of this type of management are to reduce supply-chain operating costs and to improve readiness by focusing on the customer (the warfighter) and linking demand and supply planning, purchasing, inventory management, and suppliers. The Air Force has implemented PSCM and would like to determine whether and how well it is meeting its desired improvement objectives.

Even as this initiative is being implemented, however, other factors that could affect supply-chain performance have also been changing. Yet, as our review of the literature on estimating the effects of various PSCM-type initiatives revealed, we were unable to identify any studies that explicitly account for such factors. In light of this shortcoming, we have developed a theoretical method for estimating outcomes of the PSCM initiative. This method uses an econometric model that can hold other factors constant as a means of discriminating between the changes in supply-chain performance attributable to a PSCM initiative and those attributable to other factors. AFMC supports the eLog21 program through three initiatives. The command's PSCM initiative supports the specific objective of improving equipment availability by 20 percent and reducing annual operations and support costs by 10 percent by fiscal year 2011. The means of doing so will be reducing the sourcing cycle time, reducing material purchase and repair costs, and improving supply material availability (the availability of spare parts). For purposes of illustration, we chose to focus on the last of these areas.

One indicator of supply material availability is the number of MICAP incidents. A MICAP incident occurs when a piece of equipment—an aircraft or weapon system, for example—is unable to perform at least one of its missions because it lacks a part that base supply cannot provide. MICAPs are reported at the NIIN level and are associated with the specific type of aircraft or weapon system. Theoretically, PSCM should reduce MICAP incidents by increasing material availability by improving "wholesale" response time and reducing total costs, among other things.[1] PSCM could also affect the number of parts that need to be removed because it could encourage improvements in manufacturing and repair quality and increased reliability rates. In this study, we wanted to examine how PSCM affected the number of quarterly MICAP incidents at the part level.

Our model essentially takes what economists call a *difference-in-difference* approach. It holds part- and time level effects constant, so that the estimated PSCM effect is identified by how MICAP incidents change for a given part when it is supplied under a PSCM contract. The model we describe here hints that parts repaired or purchased under PSCM contracts had fewer MICAP incidents than the same parts repaired or purchased under contracts not written under PSCM. However, the sensitivity of the results to additional explanatory variables indicates that the results must be interpreted with caution. In particular, other important, though unrelated, elements of MICAP inci-

---

[1] *Wholesale* refers to the activities conducted by AFMC and its ALCs. Decreasing administrative and production lead times and increasing contractor responsiveness would improve wholesale response times.

dents are likely to be changing concurrently with the implementation of PSCM initiatives. We hypothesize one of these to be the number of parts being removed. While some parts may be removed for maintenance on a schedule, other removals may be unscheduled because parts have failed. Among other things, the number of parts removals depends on the rate at which aircraft operate, generally referred to as *operational tempo*. As operational tempo changes over time, it will be (probably incidentally) correlated in some way with the implementation of PSCM initiatives. Because of the likely existence of these other factors, we cannot conclude that the estimated correlation between the PSCM initiative and the number of MICAP incidents represents the causal effect.

To accurately estimate the impact of PSCM initiatives on supply-chain metrics, applications of this econometric approach need to obtain data on these other factors that could be incidentally correlated with the implementation of the initiative and that affect the metrics. In our example of measuring the impact of PSCM initiatives on MICAP incidents at the part level, we use flying hours as one of these factors. However, we recommend that future applications of this econometric approach use NIIN removals as a co-occurring factor. This would permit a test of the usefulness of this approach to estimating the effect of PSCM or other initiatives on metrics that are aligned with the goals of such initiatives as eLog21.

Acknowledgments

We appreciate the help of our colleagues, Nancy Moore (the coprincipal investigator on the project) and Nancy Nicosia (a coinvestigator), who provided useful advice on organizing the material and reviewing this report. We also appreciate the helpful comments from the original program director, the late Charles Robert Roll, and other RAND analysts at an internal seminar in which we presented this work.

We would like to acknowledge the assistance of Headquarters AFMC, which provided us with contract numbers and access to the Strategic Sourcing Analysis Tool and the Multi-Echelon Resource Logistics Information Network (MERLIN) to access data. We also thank Susan Adler, RAND data librarian, who collects and archives key Air Force data used in these analyses.

We invite readers to provide comments and suggestions. Any errors and omissions are, of course, the final responsibility of the authors.

Glossary

AF/A4/7	U.S. Air Force Deputy Chief of Staff for Logistics, Installations, and Mission Support
AFMC	Air Force Materiel Command
AFMC/A4	Air Force Materiel Command, Directorate of Logistics and Sustainment
ALC	Air Logistics Center
AWP	awaiting parts
AWP incident	In an AWP incident, repair of a larger component part is delayed because one or more parts necessary for its repair are unavailable.
balanced scorecard	An approach to performance management that involves ratings taken in four key areas (the customer, finances, internal process, and learning and growth) that are then balanced with one another.
CRM	AFMC's Customer Relationship Management initiative.
commodity council	A term used to describe a cross-functional sourcing group charged with formulating a centralized purchasing strategy and establishing centralized contracts for enterprise-wide requirements for a selected commodity grouping (Reese and Hansen, 2003).

co-occurring factors	Changes occurring at the same time as the one of interest that also may bear on the desired outcome.
difference-in-difference approach	An econometric technique used to account for co-occurring factors by including entity-level and time-level fixed effects.
Depot Maintenance Transformation	AFMC's effort to reshape how the ALCs provide organic maintenance services.
DoD	Department of Defense
eLog21	Expeditionary Logistics for the 21st Century
fixed effect	An element of a model that is held constant to control for a certain category of observation, such as a time span or type of part.
FY	fiscal year
MDS	model design series
MERLIN	Multi-Echelon Resource Logistics Information Network; a data system developed for Headquarters Air Force Logistics, Installations and Mission Support for the Air Force's major commands.
MICAP	mission capable
MICAP incident	A MICAP incident occurs when a part is removed from an aircraft or weapon system and no replacement part is available from base supply, thus rendering the aircraft or weapon system unable to perform at least one of its missions.
NIIN	National Item Identification Number; the sequence of digits that uniquely identify a part

NSN	National Stock Number; a sequence of digits that describe a part. It consists of the part's Federal Supply Class (positions 1–4); NIIN (positions 5–13); and, if the part is unique to a single weapon system, its Materiel Management Aggregation Code (positions 14 and 15).
OLS	Ordinary least squares; a standard econometric technique.
operational tempo	the rate at which aircraft operate
PAF	Project AIR FORCE
Product Support Campaign	AFMC's effort to reshape how the Air Logistics Centers provide product support to already fielded systems.
PSCM	Purchasing and Supply Chain Management
Purchasing and Supply Chain Management	AFMC's effort to reshape how the Air Logistics Centers purchase goods and services from commercial companies or other government agencies and organizations.
R^2	The percentage of the variation in the dependent variable that is explained by the independent variables.
SAF/AQC	Deputy Assistant Secretary for Contracting, Office of the Assistant Secretary of the Air Force for Acquisition
SRM	AFMC's Supplier Relationship Management initiative.
strategic sourcing	Developing preferred suppliers for products or services routinely purchased from the private sector.
supply material availability	The availability of spare parts.

CHAPTER ONE

Introduction

Since 2002, Air Force Materiel Command (AFMC) has implemented best purchasing and supply-chain management (PSCM) practices at its Air Logistics Centers (ALCs) that support fielded weapon systems through the acquisition, repair, and overhaul of equipment, among other things. PSCM is the implementation of best business practices for purchasing logistics support from private-sector sources and government organizations. Its goals link to Air Force goals to improve aircraft availability and reduce the total cost of support.

The goals of the Air Force PSCM initiative are to improve supply material availability—the right parts at the right time—reduce material costs, and reduce the time it takes to provide materiel to Air Force base supply locations. Along with investments in PSCM—which have included reorganizing the way the Air Force purchases support from the private sector; training personnel to these new practices; constructing new analytical tools; and most important, writing contracts as a product of these practices—the Air Force has wanted to know if, and how much, PSCM was benefiting the warfighter. That seems logical and straightforward, but since this particular initiative was not the only one being launched at the same time, the Air Force wanted to be able to determine whether or not this particular initiative was, by itself, producing the desired results. After all, because not "all other things" were being "held constant," some of the other changes might have influenced or even been more responsible for any outcomes, good or bad. We refer to these simultaneous changes as *co-occurring factors*.

We addressed this challenge by extending an econometric model to account for these factors. This monograph describes the resulting methodology and illustrates it by applying it in a National Item Identification Number (NIIN)–level analysis of quarterly mission capable (MICAP) incidents in the context of AFMC's own PSCM initiative.[1]

This work was conducted in the context of a broad set of Air Force initiatives that the U.S. Air Force Deputy Chief of Staff for Logistics, Installations, and Mission Support (AF/A4/7) established under the umbrella of Expeditionary Logistics for the 21st Century (eLog21) to support the Expeditionary Air Force. eLog21 aims to increase equipment availability by 20 percent while decreasing annual operating and support costs by 10 percent by fiscal year (FY) 2011.[2] To meet these objectives, AF/A4/7 is supporting three AFMC transformational initiatives to reshape how the ALCs (U.S. Air Force, 2005):

- provide maintenance services, through the Depot Maintenance Transformation initiative
- purchase goods and services from commercial companies or government agencies and organizations, through the PSCM initiative
- provide product support to already fielded systems, through the Product Support Campaign initiative.

---

[1] A *MICAP incident* occurs when a part is removed from an aircraft or weapon system, whether because of unexpected failure or scheduled maintenance, and no replacement part is available from base supply, thus rendering the aircraft or weapon system unable to perform at least one of its missions.

The NIIN is part of a longer number that the Department of Defense (DoD) assigns each commercial part for material management purposes. That longer number is known as the National Stock Number (NSN). Its first four digits indicate the Federal Supply Class (also called the Federal Commodity Class), which can vary over time as parts are reclassified. The next nine digits, the NIIN itself, identify the specific part and therefore typically do not vary. The last two NSN digits are the Materiel Management Activity Code, which indicates the aircraft or engine type. Common items that are used on more than one aircraft type and items managed by the Defense Logistics Agency do not have this code.

[2] eLog21 is the Air Force Logistics Transformation initiative that supports DoD's joint vision of a modern expeditionary Air Force (Roche and Jumper, 2005).

We focus here on the second of these. Many companies have recognized the importance of the contributions their suppliers make to core business operations. PSCM was born out of the recognition of the need to integrate the acquisition of such resources, such as repair and spare parts, into supply-chain operations. A growing body of literature has shown how innovative companies are identifying and applying best practices for purchasing and for managing their suppliers, supply bases, and supply chains (Moore et al., 2002). AFMC has adapted these best practices through its own PSCM initiative.

Generally, the objectives of best purchasing and supply management practices are to lower supply-chain costs and improve performance by

1. rationalizing contracts and suppliers for related goods and services, which often means substantially reducing the numbers of contracts and suppliers
2. selecting the best suppliers with the lowest total cost of ownership, that is, those with the best technology, highest quality, best delivery, or lowest price
3. developing strategic relationships with key suppliers
4. working with key suppliers on continuous improvements.

The AFMC PSCM initiative, established to support eLog21 goals, has several of its own objectives to meet by FY 2011:

1. reducing sourcing cycle time by 50 percent[3]
2. improving supply materiel availability by 20 percent[4]
3. decreasing materiel purchase and repair costs by 20 percent.

PSCM also aims to increase the time on wing and the mean time between failure rates for individual parts (Dryden and Tinka, 2004).[5]

---

[3] *Sourcing cycle time* is the time it takes from order to delivery.

[4] *Supply materiel availability* measures whether the wholesale level can fill a requisition for a stocked item.

[5] *Time on wing* refers to the mean time of operation between engine or part removals. The Multi-Echelon Resource and Logistics Information Network (MERLIN) data system

Air Force Materiel Command PSCM Initiative

In March 2002, AFMC began its PSCM initiative at Oklahoma City ALC and, in April 2003, began implementing it enterprisewide (AFMC, 2004). AFMC's PSCM initiative has four major components, all based on commercial best practices (Gabreski, 2004):

1. commodity councils
2. customer relationship management
3. supplier relationship management
4. balanced scorecards.

The Air Force has reorganized its formerly decentralized ALCs' component-purchasing activities into eight *commodity councils* that develop supply strategies for groups of similar purchased sustainment goods and services for AFMC (U.S. Air Force, 2006). By centralizing its sustainment NIIN–related requirements across the command, the commodity councils are able to leverage AFMC's entire sustainment business for these items with suppliers and can negotiate more-favorable terms by consolidating sole-source purchases into fewer contracts, known as *corporate contracting*. It also seeks to contract with preferred suppliers for products or services routinely purchased from the private sector, known as *strategic sourcing*.[6]

The *Customer Relationship Management* initiative (CRM) provides a single customer point of contact for all aspects of materiel management (AFMC, 2006). Its incarnation at AFMC will make use of the Expeditionary Combat Support System, the Air Force's enterprise resource planning information system, when it is fully developed. Customer relationship management will provide a single, consistent interface that provides accountability for serving the customer.

---

defines *time on wing* as flying (or operating) hours divided by the sum of scheduled and unscheduled engine removals. At the NIIN level, it is the flying (or operating) hours divided by scheduled and unscheduled part removals from an engine.

[6] The Office of Management and Budget also set similar goals in a May 2005 directive (Johnson, 2005).

The *Supplier Relationship Management* (SRM) initiative establishes teams led by senior executives to work with suppliers with whom the Air Force spends the most. Each executive manages and develops suppliers to continuously improve their costs and performance year after year and helps the Air Force become a better customer. It is currently established for sustainment spending, although the Air Force has plans to implement it elsewhere as well.

The *balanced scorecard* is based on concepts developed by Kaplan and Norton (1992). It is designed to measure performance based on four perspectives: financial, customer, internal process, and learning and growth. This method has spawned a wide set of analyses, especially within the military or from military contractors (e.g., Kem et al., 2000; Gorski, 2005). AFMC regularly reviews these corporate scorecards and uses them to identify problem areas to address. It also consults these scores when evaluating past performance.

Is PSCM Helping AFMC Reach eLog21 Goals?

For the Air Force, the ultimate question here is whether or not the PSCM efforts just described are in fact helping AFMC achieve its eLog21 goals and, if so, by how much. In addition, is the Air Force seeing a return on its PSCM investments? RAND Project AIR FORCE was asked to develop a methodology that could measure the benefits of PSCM initiatives and would focus on performance improvements.

The difficulty in measuring the benefits of initiatives stems from the fact that during the implementation of PSCM, many other dynamic environmental factors besides materiel support have been influencing operational outcomes—flying hours, other initiatives such as Depot Maintenance Transformation, problems unique to given weapon systems, etc. So, the primary challenge in estimating the benefits of PSCM initiatives would be to separate out factors other than PSCM that were also affecting the supply-chain and operational outcomes. We refer to these as *co-occurring factors*, defining them as factors that may affect purchasing and supply-chain management efficiency (and certain metrics) and that may be correlated with the implementation of a PSCM initiative. Note that we do not simply denote them as "explanatory variables" (the typical term for variables available for the econometric model described later) because they are not necessarily observable or available, and we want to stress that these factors change at the same time that PSCM implementation occurs, even though they are theoretically unrelated. Given the inevitable existence of co-occurring factors, isolating the effects of the PSCM initiative on operational outcomes is quite complex. A simple "before and after" comparison, which is the typical approach we found in the military literature, is inadequate for answering the Air Force's question.

This monograph describes a method that attempts to isolate the particular, causal effect of PSCM on operational outcomes using econometric models. Ideally, such models would estimate the effects of the implementation of the PSCM initiatives on the outcomes PSCM initiatives target (such as sourcing cycle time or material availability). In the end, however, the usefulness of the model in isolating the causal effects of the PSCM initiative will rest on being able to adequately control for co-occurring factors.

Testing the Model

To test our model, we analyzed a set of NIINs from an initial set of 28 contracts awarded under AFMC's PSCM initiative as of March 2005.[7]

We began by determining the number of quarterly mission capable (MICAP) incidents particular to each NIIN in our data sample set. The number of MICAP incidents is indicative of material availability and is related to aircraft mission capability and therefore aircraft availability, which is one of the primary performance measures among

[7] Only two contracts had been written under PSCM at the time of this analysis, so we also included corporate and strategic sourcing contracts, because they share some of the same characteristics of a PSCM contract. They are usually long term and cover many NIINs.

eLog21's objectives.[8] PSCM initiatives could reduce the number of MICAP incidents by improving the availability of parts.

The co-occurring factors we found for aircraft-related NIINs included the number of flying hours for the associated aircraft. We were, however, unable to obtain co-occurring factors for such other NIINs as those related to engines, as we discuss in Chapter Four. As it turns out, our estimates of the effects of PSCM initiatives changed with the inclusion or exclusion of the time-level fixed effects, NIIN-level fixed effects, and (for aircraft-related NIINs) flying hours.[9] This suggests that the results must be interpreted with caution. Nevertheless, the econometric model could serve as a reference for future attempts at similar evaluations.

Organization of This Monograph

Chapter Two describes the importance of accounting for co-occurring factors and provides a review of the literature on tracking metrics and estimating the impact of initiatives similar to PSCM. In Chapter Three, we describe how an econometric model can address the problem of co-occurring factors, develop a general econometric model, and discuss empirical issues for applying the econometric model. We describe an application of this model to an analysis of MICAP incidents in Chapter Four. We present our results in Chapter Five. In Chapter Six, we discuss our conclusions and implications.

---

[8] Note that mission capability is affected by MICAP incidents, as well as incidents affecting base maintenance availability.

[9] As we explain in Chapter Three, time-level fixed effects are average levels of the outcome over time, holding constant the influences of other factors. Including time-level fixed effects makes the values of the other variables deviations from the mean for each period. Likewise, NIIN-level fixed effects are average levels of the outcome across NIINs, holding constant the other factors, and including NIIN-level fixed effects makes the values of other variables deviations from the mean for each NIIN.

CHAPTER TWO

Background and Motivation

The Importance of Accounting for Co-Occurring Factors

The primary challenge in isolating the causal relationship between PSCM and supply-chain and performance outcomes is adequately controlling for co-occurring factors. Co-occurring factors are those that affect the outcomes and are correlated with the initiative, either incidentally or by design, and happen to occur concurrently with the initiative. Distinguishing the effects of the initiative on outcomes of interest *from* the effects of the co-occurring factor(s) that are also affecting the same outcomes of interest requires accounting for all relevant co-occurring factors.

As an example, suppose that the Air Force implemented a new supply-chain initiative in a given year. Also suppose that the United States becomes engaged in extensive military operations in that year, requiring increased Air Force flying hours. The increase in Air Force flying hours could potentially create greater demands on the supply chain, which could negatively affect supply-chain metrics. Because this could result in an understatement of the benefits of the supply-chain initiatives, the Air Force might not gain a good assessment of the actual effects of a particular initiative.

As a notional example of the effects of co-occurring factors, consider Figure 2.1, which shows the change from one period to the next in a metric (the number of MICAP incidents) and in co-occurring factors. The rectangular "events" can be observed, while the oval "effects" are unobserved and unknown. The Air Force wants to estimate the

Figure 2.1
A Notional Example for a Change: Before and After the PSCM Initiative

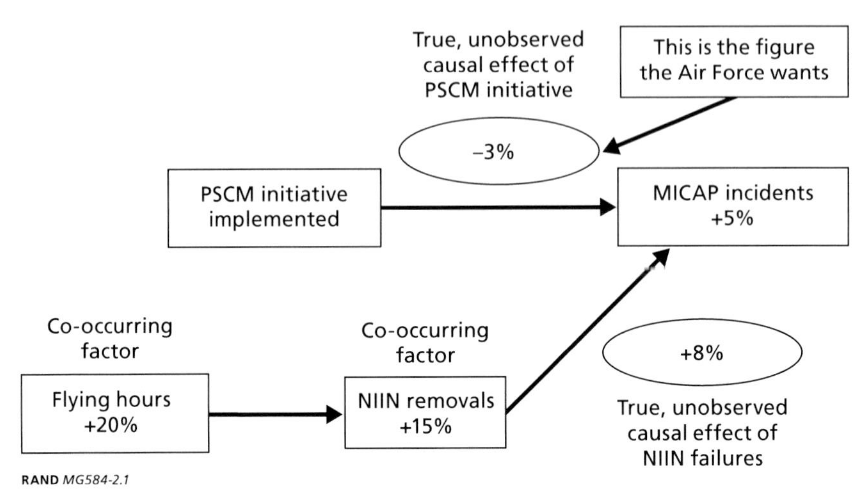

RAND *MG584-2.1*

true effect of the PSCM initiative on MICAP incidents. Let us suppose that the true (unobserved) effect is a 3-percent reduction in MICAP incidents due to improved material availability. In this case, the co-occurring factor is NIIN removals, which was caused by what can also be considered a co-occurring factor: the amount of flying hours. In this example, the amount of flying hours increases by 20 percent, which leads to a 15-percent increase in the number of NIIN removals. While we would not observe it, let us say that these NIIN removals cause an 8-percent increase in the number of MICAP incidents.

As we describe below, the typical approach to examining the effects of initiatives is a before-and-after comparison of the metrics, without taking into account any other factors that may also have changed. In our example, the traditional approach would show that the PSCM initiative was associated with a 5-percent increase in the number of MICAP incidents. The 5-percent increase would be the sum of the effects from the PSCM initiative (–3 percent) and the additional NIIN removals (+8 percent). However, because these two individual effects are occurring at the same time and are unknown, the only number the evaluator observes is the 5-percent increase. So, because these co-occurring factors affecting outcomes in their own way are not properly accounted for, the effect from changes in NIIN removals is inappropriately allocated to PSCM. Thus, in this example, this approach would understate the causal effect of the PSCM initiative on MICAP incidents and could even suggest that the PSCM effect had been negative. This could cause the Air Force to underestimate the effects of PSCM and could affect future Air Force resource decisions. In addition, this could provide incorrect feedback on the efforts of personnel who are actually bringing benefit to the Air Force, because, in our notional example, without their efforts in implementing the initiatives, things would have been worse.

If, on the other hand, fewer parts were removed at the same time the PSCM initiative was implemented, there could be fewer MICAP incidents from both the initiative and the decrease in removals. Thus, we would likely see the opposite result from above—that the estimated effects of the PSCM initiative would overstate its positive benefits.

To gain a better estimate of the true causal effect of PSCM initiatives, analysts need to account for relevant co-occurring factors that could also be affecting metrics.

Literature Review

We reviewed the academic, trade, and defense literature on developing methods to measure the effects of purchasing initiatives on supply-chain performance and whether performance goals were being met. While we identified literature on supply-chain metrics, we found little reported on developing the empirical basis for using these metrics.

Most studies on performance metrics used the balanced scorecard approach. Balanced scorecards can be used for tracking performance over time. Thus, the balanced scorecard is useful for point-in-time comparisons of the performance of different entities. But estimating the benefits of PSCM requires a method that can account for factors other than supply-chain performance that can affect metrics (such as the level of overall business), and the balanced scorecard has no formula or model that accounts for other factors that may affect the metrics being measured. Still, the inputs for the balanced scorecard could potentially also be useful for a model such as the one we have developed.[1]

For the most part, the literature on the balanced scorecard and related topics describes the development of metrics to measure the performance of the supply-chain and purchasing strategies; examples include Stewart (1995) and Gunasekaran, Patel, and Tirtiroglu (2001). We did not identify any literature, however, that seems to address the problem of disentangling the effects of purchasing initiatives from those of other factors affecting performance outcomes.

Some studies have specifically focused on particular companies to demonstrate how the implementation of certain PSCM initiatives changed certain purchasing metrics. The approach commonly used in these studies is to track how metrics change over time or to compare the metrics from before and after the purchasing or supply-chain initiatives. However, as with the balanced scorecard studies, other factors, unrelated to the initiatives, may also have changed over periods of observation that might also have caused changes in the metrics. For before-and-after comparisons, the difference may partly reflect the effects of the co-occurring factors, which would mean that the estimated causal effect of PSCM initiatives on the measured outcomes would be incorrectly estimated. This could lead decisionmakers to withhold resources from initiatives that are beneficial or divert resources to initiatives that may not be beneficial.

An example of potentially misleading indicators of change in PSCM efficiency comes from the journal *Purchasing*. Avery (2003) described purchasing and supply-chain initiatives that Rockwell Collins, a company that provides communications and aviation electronics equipment to commercial and military customers, had implemented:

> As a result of efforts of the enterprise sourcing team over the past 18 months, Rockwell Collins has reduced purchasing costs up to

[1] Our discussions with members of the Headquarters AFMC Directorate of Logistics and Sustainment, Supply Operations Division, Supply Policy and Analysis Branch on November 15, 2006, confirmed that the balanced scorecard tracks metrics over time and does not internally account for co-occurring factors that could affect the observed metrics.

20% of some commodities, consolidated the supplier base and improved lead-times 30%.

However, the period she refers to is mid-2001 to late 2002, which was also a time of decreasing demand for aerospace products. In fact, Rockwell Collins sales dropped 12 percent from FY 2001 to 2002, which could, though not necessarily, also have affected purchasing costs.[2] Fewer sales could reduce stress on the supply chain, which could improve lead times, regardless of any PSCM initiatives. Thus, it is difficult to ascertain the initiative's causal relationship with these metrics without accounting for other large changes occurring coincidentally.

A later article on Rockwell Collins had potentially opposite implications. Avery (2005) notes that further focus on improving supply-chain management led to a decrease in the average lead time from 46 days in 2002 to 30 days in 2005 and to improvement in on-time deliveries from 83.8 percent in 2002 to 96.5 percent in 2005. These statistics are potentially more significant than one might initially think, especially when considering that the business grew significantly since 2002: Revenues increased 38 percent from FY 2002 to FY 2005 (see Rockwell Collins, Inc., 2005).

---

[2] The sales statistic comes from Rockwell Collins, Inc. (2002). The revenue growth rate indicated here and below could be affected by a few acquisitions and divestitures that occurred during these periods. However, with these acquisitions and divestitures, the company could be gaining or losing part of its supply base. It is not possible to determine from the annual reports how much of the changes in the supply-chain statistics are attributable to these acquisitions and divestitures.

CHAPTER THREE

Model Methodology and Data Requirements

Applying an Econometric Model to Estimate the Benefits of PSCM Initiatives

The goal of this particular analysis was to estimate the effect of the implementation of PSCM on the number of MICAP incidents. The challenge, as described in the previous chapter, was to control for all relevant co-occurring factors.

Several techniques have been developed to estimate the effect of one factor on a particular outcome. The typical approach is an econometric model. Such a model can be estimated for outcomes at various entity levels, such as NIINs, weapon systems, and an enterprise (such as, in this case, the Air Force as a whole). As we explain later, this type of model is more suited for lower-level entities, such as NIINs. An econometric model tries to determine the relationship between several explanatory variables (in our case, implementation of PSCM initiatives and co-occurring factors) and a dependent variable (the outcome). The coefficient estimates indicate how much the outcome changes with each unit change in the independent variable, assuming that all other variables are held constant.[1] Theoretically, regression models can estimate the causal effects of the PSCM initiatives on supply-chain outcomes more accurately by holding co-occurring factors constant, so that their effects can largely be separated from the estimated effect of

[1] Typically, such a model finds the relationship between these variables (the coefficient estimates on the explanatory variables) that minimizes the difference or the squared difference between the actual values and the predicted values based on the coefficient estimates.

the PSCM initiative. However, this requires that adequate data for the co-occurring factors are available.

What makes this difficult is that, because PSCM is being applied to Air Force NIINs gradually over time, the initiative's effects will inevitably be correlated in some way with many other factors that also change over time. In the notional example we provided in the previous chapter, we showed that changes in operational tempo (and thus, part removals) could be correlated with PSCM implementation, so that the estimated effect of the PSCM initiative would partly reflect the effects of part removals on MICAP incidents if the model were not able to control for part removals.[2]

Without fully controlling for all major relevant factors, it is not possible to rule out these other factors as causes of the changes between pre- and post-implementation. Table 3.1 shows examples of factors of MICAP incidents, categorized as those external and those internal to the supply chain. Two factors external to the supply chain are the number of removals for the NIIN and scheduled removals. Scheduled removals based on calendar time should be constant over time for

Table 3.1
Examples of Factors Influencing the Number of MICAP Incidents for a Particular NIIN

Supply-Chain Factors	Sources of Variation
External	
Unscheduled removals (part failures)	Operational tempo and age of parts
Scheduled removals	Time, operational tempo, and sorties
Internal	
Stock-leveling policies	Funds and total inventory
Repair capacity	Funds and anticipated business volume

[2] In our example, we showed that under PSCM implementation, flying hours and part removals increased independently of PSCM, though concurrently. If flying hours or part removals were not included in the model, then the effect of the PSCM initiative would erroneously be estimated as negative, even though in our example it was positive.

the NIIN, but other scheduled removals may be based on operational tempo and the number of sorties. Two factors among many that are internal to the supply chain are the methods for setting stock levels for the NIIN and repair capacities at bases and depots. To the extent that these change over time, they would need to be controlled for as well. However, they should not be controlled for if they are a product of the PSCM initiative because that would prevent the model from estimating the full effect of the PSCM initiative.

It is possible that several kinds of PSCM initiatives could be implemented at the same time. In this situation, isolating the causal effects of one particular initiative would require controlling for all other initiatives as well. Unfortunately, many of these co-occurring factors are not readily available or are difficult to quantify for a regression model. Economists have developed well-established techniques for reducing what is known as *omitted variables bias*, which is the technical term for the problems associated with being unable to control for these co-occurring factors.

One technique that could be applied in this situation is to use a difference-in-difference approach, which entails the inclusion of entity-level and time-level fixed effects. Fixed effects essentially hold constant some factor for a set of observations. Entity-level (e.g., NIIN-level, or what is being modeled) fixed effects would hold constant the entity, so that differences in the number of MICAP incidents across entities would be accounted for and not reflected in the estimated effect of PSCM. Likewise, time-level fixed effects hold constant a particular period, thus accounting for factors that influence the number of MICAP incidents for all entities. This model would base the estimated effect of the PSCM initiative on the average within-entity differences in the metric given a change in PSCM status.

Still, it is likely that some factors will affect certain entities (such as the weapons that are playing the central role of a particular operation or exercise) and not others. And such factors would be correlated with PSCM implementation unless the average influence of these factors is the same before and after PSCM implementation for the NIIN. Thus, the time- and entity-level fixed effects will probably not fully account for the relevant co-occurring factors. Therefore, efforts need to be made to account for these co-occurring factors that vary with time and that are specific to certain entities being modeled, such as NIINs.

Applying the Econometric Model for NIIN-Level Outcomes

Econometric models are well suited for examining metrics at the NIIN level when the timing of PSCM implementation and measurements for observed metrics for NIINs vary. NIINs can differ from each other in many ways, including complexity, cost, usage (that is, by the end items associated with the NIINs), and mission essentiality. NIINs may also differ in their general removal rates. For these reasons, stocking policies for particular groups of NIINs may differ, which could affect such metrics as MICAP incidents. In addition, the NIINs that are targeted for PSCM may be those associated with the worst outcomes, such as high MICAP rates. An implication of these differences across NIINs is that one cannot just compare, at a given moment, the metrics for PSCM NIINs against those for non-PSCM NIINs because differences in the metrics could be due to inherent differences between these NIINs. In that case, the estimated effect of the PSCM initiative could partially reflect differences across NIINs. For example, if, hypothetically, NIINs that were more difficult to support and had worse metrics were the ones targeted for PSCM initiatives, then a simple comparison of PSCM NIINs to non-PSCM NIINs would likely associate worse metrics with being part of the PSCM initiative.

Thus, including a set of NIIN-level fixed effects is essential for preventing the estimated PSCM effect on outcomes from reflecting key differences across NIINs. Including NIIN-level fixed effects essentially adds a set of NIIN dummy variables so that the identification of the effect of the PSCM initiative will come from within-NIIN changes in the outcome between before and after the PSCM initiative is implemented for the given NIIN, holding constant other factors. Thus, the identification of the PSCM effect will not rely at all on differences across NIINs.[3]

Incorporating the NIIN- and time-level fixed effects, the econometric model would have the following general form:

$$Y_{kt} = \mu_k + \mu_t + \gamma\left(PSCM_{kt}\right) + \lambda\left(C_{kt}\right) + \varepsilon_{kt}, \tag{3.1}$$

where

Y_{kt} = a measure of supply chain performance (number of MICAP incidents)

k = NIIN

t = time (quarter)

μ_k = NIIN-level fixed effect

μ_t = time-level fixed effect

$PSCM_{kt}$ = indicator for whether NIIN k was on a PSCM contract at time t

C_{kt} = a vector of co-occurring factors

ε_{kt} = error term

γ = the coefficient estimate representing the effects of PSCM

λ = the coefficient estimate(s) on the vector of co-occurring factors.

---

[3] In the economics literature, fixed effects are commonly used when there are repeated observations for a given entity. For example, in analyses on how the economy (specifically, the state unemployment rate) affects substance use, Ruhm (1995) and Arkes (2007) include state fixed effects to make sure that incidental correlation between inherent differences across states in the unemployment rate and in the level of substance use do not affect the estimated impact of the unemployment rate. With the state fixed effects, the identification of the effect of the unemployment rate comes from within-state changes over time in the unemployment rate and substance use.

This model estimates how PSCM and the co-occurring factors (*C*) affect the outcome, while holding constant the specific NIIN and the quarter. It thus examines the within-NIIN changes over time in the outcome, relative to other NIINs, attributable to the within-NIIN change in the status of whether the NIIN is under a PSCM contract.[4] We should note that we are following a common application of econometric models that has been used in many economic analyses.[5]

Given its wide applicability, our model has applications in the Air Force beyond estimating the effects of PSCM initiatives. For example, with adequate data, the model could estimate the effects of certain initiatives or environmental factors on customer wait time for particular parts. The various levels (corroborating to NIINs in our model) could be different Air Force bases.

Note that the model we specify in Equation 3.1 is a generic form of an econometric model. The researcher must choose a functional form for the model based on the distribution of the dependent variable. The most common ones are ordinary least squares (OLS) models for outcomes that approximate a normal (bell-shaped) distribution and probit or logit models for dichotomous outcomes. However, a recent trend has been also to use OLS for dichotomous outcomes as long as the average value of the variable is not too close to zero or one.

Empirical Issues with Econometric Models for NIIN-Level Analyses

Econometric models at the NIIN observational level have three important empirical issues that ought to be addressed. The first is whether and how to weight observations (in our case, NIINs). If the Air Force wanted to take into account a dimension that gives particular NIINs

---

[4] See Greene (2002) for more details on econometric models.

[5] For example, Arkes and Kilburn (2005) analyzed the influence of various factors (such as the unemployment rate and public-school tuition costs) on active-duty and reserve enlistments. They used state (of origin) and year fixed effects to control for the effects of inherent differences across states in propensity to enlist and trends over time that may be incidentally correlated with these other factors.

greater importance (perhaps mission essentiality or cost), a methodology for how to weight observations (NIINs) would need to be developed. These weights would then be applied to the economic model so that the estimated effects would be based on important criteria related to NIIN differences that the analyst wants to represent.

The second issue concerns interpreting results. As discussed earlier, there could be differences across observations (NIINs) in several dimensions (such as usage and cost for NIINs). One of these dimensions may be in how much a PSCM initiative can improve metrics for the NIINs. For example, the Air Force may have targeted NIINs that have metric outcomes it especially wanted to improve or NIINs that it believes would have received the most benefit (with the greatest improvement in outcome metrics) from the initiative. Thus, improvements in metrics for a certain set of NIINs may be specific to those NIINs and different from potential improvements for a different set of NIINs.

A third issue is which observations (NIINs) to include in the analysis. The estimated effect of the PSCM initiative (the coefficient estimate for γ) is identified by the variation in the timing of the implementation of the initiatives for the NIINs associated with a PSCM contract. The estimated PSCM effect will also depend on the estimated time-level fixed effects and the effects of the co-occurring factors. Thus, it is important to estimate these effects accurately. Using the larger set of all Air Force NIINs, PSCM and non-PSCM alike, would make estimates of the time-level fixed effects and the effects of the co-occurring factors more accurate. The greater accuracy would come from having a larger, and perhaps more representative, sample of NIINs. On the other hand, if the NIINs with transformation (in our case, the PSCM implementation) are a distinct set of NIINs and are different from the average NIIN, that would justify limiting the analysis to the transformed NIINs. The results of these two approaches are likely to be different, and there is no clear answer, in our situation, as to which one of these approaches is better. Our strategy in the next chapter is to estimate the model with just the PSCM NIINs.

Difficulties in Estimating a Model for Enterprise- and Weapon-Level Metrics

In some situations, the question may be how a certain initiative affects metrics for a particular weapon system or for the Air Force as a whole. In this case, the model would need to be estimated at the weapon or enterprise level. Unfortunately, the econometric model in Equation 3.1 is difficult to operationalize at the enterprise or weapon level, such as for net operating results or aircraft availability for a specific weapon system. At the enterprise level, there would not be any variation in the timing of the implementation of the PSCM initiative. That is, the model would have only one entity (the whole enterprise) and one observation per period, rather than separate entities (NIINs) and multiple observations per period. Thus, the model would be reduced to the following:

$$Y_t = \gamma\left(PSCM_t\right) + \lambda\left(C_t\right) + \varepsilon_t. \tag{3.2}$$

At the enterprise level, with the single-entity model, the unit of observation is only a period of time, as opposed to the unit of observation being a NIIN period in Equation 3.1. Consequently, the model has no NIIN subscript. The primary concern with this approach is that it would not be possible to include time-level fixed effects because the timing of the PSCM initiation would correlate perfectly with the time variables because no multiple entities vary in the timing of implementation.

A single-entity-model approach could work if an analyst could fully control for relevant co-occurring factors, such as the total number of NIIN removals across the Air Force, for an analysis on MICAP incidents. Otherwise, time-level effects would have to be represented with a linear or quadratic trend.[6] However, if the co-occurring factors do

---

[6] The trend terms in an econometric model factor out time effects but are constrained to be of a certain shape. A linear trend has a constant slope over time. A quadratic trend allows a nonlinear shape, with no inflection point.

not follow a linear or quadratic pattern, these terms will not adequately represent the factors.

Taking a weapon-level approach (e.g., aircraft availability) is also difficult. First, examining only one weapon at a time would lead to problems similar to those for an enterprise-level analysis because there would be no variation in the timing of the PSCM implementation. Thus, it would be difficult to distinguish the PSCM-level effect from other factors that change over time. Second, PSCM initiatives have been typically targeted at the NIIN level, and each weapon system has many NIINs. Thus, the relationship between a given NIIN on a weapon system and the metrics for the weapon system itself may be weak and highly diffused. That is, a handful of NIINs subject to PSCM implementation may not materially affect metrics at the weapon level. On the other hand, PSCM contracts may be formulated to improve the supply chain for certain NIINs that strongly affect metrics for a specific weapon system. In this case, the second issue of a weak relationship between the NIIN and weapon-level metrics would not be as relevant. However, the first issue of no variation in the timing of the implementation of PSCM would still exist.

CHAPTER FOUR

Applying the Model to an Examination of MICAP Incidents

This chapter presents our illustrative analysis using the framework just described. Our primary outcome (for the first set of models) is the number of quarterly MICAP incidents for a NIIN. A secondary outcome is a dichotomous (0 or 1) variable for whether *any* MICAP incidents occurred in a given quarter for a particular NIIN.

The PSCM effort we examined included 26 corporate and strategic sourcing contracts and two PSCM contracts.[1] Of these 28 contracts, 17 satisfied our sample criteria of (1) becoming active just before the start of FY 2001 or later and (2) having purchases in the NIIN-level contract data for Air Force–managed items between FY 2001 and FY 2004.[2] For our analysis period, the value of these contracts ranged from $27,000 to 518 million, and 1 to 158 spare or repair NIINs were purchased. Overall, our sample had 624 different NIINs purchased under the 17 contracts. The dates of implementation of the PSCM contracts, measured by the first time they (or the NIINs) were observed in

[1] We received these contract numbers from AFMC's Supplier Management Division in May 2005.

[2] We did not analyze contract data for the Defense Logistics Agency. Among the excluded Air Force contracts was one that began in February 2000. Because our contract data began in FY 2001, we could not determine when the NIINs on this contract were first supplied under a PSCM contract. We did include another contract that began in mid-September 2000 that had 22 NIINs in the sample. At least one-half of the NIINs were observed in the detailed FY 2001 data and so would be off by no more than one year for the implementation.

contract data, ranged from mid-September 2000 to August 2004. The appendix provides detailed information on these 17 contracts.[3]

To capture data on MICAP incidents, we established the quarter as the basic unit of time and set the overall period for our assessment at 25 quarters, ranging from the first quarter of FY 1999 to the first quarter of FY 2005. We chose this period, which includes the two years before the implementation of any PSCM contracts, to ensure that our baseline measures for the number of MICAP incidents were adequate. Then, for each of the NIINs in our sample, we counted the number of MICAP incidents that occurred in each quarter of our assessment period.[4] We collected a total of 15,600 observations: 25 quarters for the 624 NIINs.

Data Samples

We first analyzed the 15,600 observations, representing the 624 NIINs over 25 quarters, then separated the NIINs according to the type of end item (aircraft versus engine) each was most frequently reported to be on and the model design series (MDS) data element reported in MICAP data. This required us to observe the NIIN in the MICAP data. Thus, we created a sample of the 357 NIINs that had had at least one MICAP incident in the 25 quarters. While including all 624 PSCM NIINs would be more technically correct, the latter sample was

[3] Our contract data came from the Global Combat Support System—Air Force (formerly Air Force Knowledge Service) Strategic Sourcing Analysis Tool, which AFMC developed to support its commodity councils. Our MICAP data came from the MICAP Analysis and Reporting Tool, which AFMC Directorate of Logistics and Sustainment, Supply and Engineering Division developed and maintains. Flying hour data are from MERLIN, a data system developed for Headquarters Air Force Logistics, Installations, and Mission Support for the Air Force's major commands (U.S. Air Force, 2003).

[4] As a point of comparison, during the period we examined, 2.14 million MICAP incidents occurred across the Air Force. These incidents involved 228,389 different NIINs, of which 357 involved a PSCM contract.

more meaningful for comparison to the separate aircraft and engine samples. The result was 128 aircraft NIINs and 146 engine NIINs.[5]

We also considered estimating the model with a more-universal set of NIINs. Theoretically, this would provide a more-accurate set of time-level fixed effects for the model because they would be for the Air Force rather than only the initial set of PSCM NIINs. Unfortunately, the only set of NIINs that we would be able to use would be the set of 228,389 NIINs that had at least one MICAP incident in our 25-quarter time frame. Thus, this sample would not be representative of the whole Air Force. Furthermore, in such a model, only about 0.1 percent of the NIIN-quarter observations would fall under a PSCM contract.

Co-Occurring Factors

As mentioned in Chapter Three, the ideal co-occurring factor would be the number of NIIN removals. However, data on NIIN removals were not readily available.[6] Thus, we sought a factor that is associated with the number of removals. For aircraft NIINs, we used the number of flying hours for the end item (the MDS) that is most often associated with the NIIN. Flying hours serve as a proxy for unscheduled NIIN removals. These have been used in many previous studies on predicting part failure rates (for example, Adams, Abell, and Isaacson 1993; Hillestad, 1982). Furthermore, flying hours are part of the product select codes that are used to forecast future failures in the Air Force's requirements determination models. These codes include flying hours, inventory, sorties, drone recoveries, and ammunition expenditures.

Flying hours are generally available, but this variable has limitations. First, flying hours are less directly linked to MICAP incidents

[5] These numbers do not add up to 357 because two of the PSCM NIINs were not related to aircraft or engine systems, and 81 could not be linked to any system.

[6] Base maintenance personnel record removals of work unit codes, which are related but not identical to NIIN failures. D200A, which is part of the Air Force's Secondary Item Requirement System, contains data inputs to the requirements model NIIN demand data report rates, which are smoothed over multiple quarters (AFMC, 2005).

than are NIIN removals, so they would likely explain much less of the variation in MICAP incidents.[7] In addition, a NIIN may be common to more than one weapon system, so it would not be clear which system's flying hours to assign to a given NIIN. And some NIINs may not be linked to weapon systems for which information on flying hours is available (such as ground-based radars).[8] Furthermore, we were unable to obtain flying hours for some airframe-related NIINs because our data source (MERLIN) did not report flying hours for a few airframes. For example, even though there were many MICAP incidents for NIINs on H-53s, MERLIN did not report flying hours from the first quarter of FY 2001 onward.

Operationally, we identified the MDS most often associated with the NIIN in the MICAP data. We then assigned it to the higher-level model design and summed up the monthly flying hours for the MDSs under a given model design. We did this to be consistent and maximize our data set because several MDSs did not have information on flying hours.[9] We wanted to use analogous information for engines but were unable to obtain the necessary data. The commonly used statistic for engines is engine time on wing, which is defined as engine operating hours divided by the sum of scheduled removals and unscheduled removals. Because the effort required to obtain data on removals exceeded project resources, we were unable to derive data on engine operating hours. Thus, the only variables that could be considered co-occurring factors (C_{kt}) for the sample of all NIINs and for the sample of engine NIINs are the time-level fixed effects.

---

[7] Flying hours are the same across all aircraft NIINs for a particular MDS, whereas part failures are NIIN-specific. Other product select codes factors, such as sorties, share the same characteristic as flying hours, i.e., they apply to all MDS NIINs equally.

[8] Scheduled removals of some NIINs are related to criteria other than flying hours, such as operating time or calendar time. Collecting such data was beyond the scope our work.

[9] For example, the MICAP data for the EC-135 include many MDSs: EC-135A, C, E, H, J, K, N, and P. However, flying hours were available only for the EC-135K and EC-135N. Generally, the MDS flying hours cover most of the large MDSs related to a given model design. However, aggregating the MDSs up to the model design level allows more completeness.

Sample Results

Table 4.1 shows the descriptive statistics we calculated for the four samples. The first is the sample based on the 624 PSCM NIINs observed in the contract data. The second, third, and fourth samples are based just on NIINs that had at least one MICAP incident in the 25 quarters we examined. The second sample has observations for all 357 of these, while the third includes observations only for the 146 engines NIINs and the fourth only for the 128 aircraft NIINs. The resulting sample sizes are noted in the table. From FY 1999 to the first quarter of FY 2005, the average number of MICAP incidents per quarter was 0.92 for all NIINs and 1.62 for those with at least one MICAP incident. The average was slightly higher for aircraft NIINs and even higher for engine NIINs. The same pattern applies for the probability of at least one MICAP incident occurring in a given quarter. For all NIINs, 37 percent of the observations occurred after the PSCM implementation for the specific NIIN—indicated as "Was under a PSCM contract." The percentages are similar for the other three samples.

For these samples, we applied a form of Equation 3.1 from above to two outcome measures: the number of MICAP incidents for NIIN, k, in quarter, t, and a dichotomous indicator for whether the NIIN had any MICAP incidents in a given quarter. For the separate aircraft models, the NIIN-specific co-occurring factor is the number of flying hours. Flying hours were missing for 1 percent of the NIIN-quarter observations. In these cases, following a standard technique for dealing with missing values, we assigned zero to the flying hours and a value of one to the NIIN for a missing-hours indicator variable.

As discussed above, Equation 3.1 is the generic econometric model for which we must specify a type of econometric model based on the distribution of the outcome metric. Figure 4.1 shows the population density function (the probability distribution of various values of a given variable) of the number of quarterly MICAP incidents (the dependent variable) for the second sample of all PSCM NIINs with at least one MICAP incident. For comparison purposes, it also shows the discrete normal population density function as a bell curve, assuming the same mean and standard deviation for the actual data. Significant

Table 4.1
Descriptive Statistics for the Sample of PSCM NIINs

In One NIIN-Quarter	Number	Mean	Standard Deviation
All PSCM NIINs			
NIINs	624		
Observations	15,600		
MICAP incidents		0.92	3.10
Had at least one MICAP incident		0.22	0.41
Was under a PSCM contract		0.37	0.48
All PSCM NIINs with at least one MICAP incident			
NIINs	357		
Observations	8,925		
MICAP incidents		1.62	3.96
Had at least one MICAP incident		0.38	0.49
Was under a PSCM contract		0.34	0.48
Aircraft NIINs			
NIINs	128		
Observations	3,200		
MICAP incidents		1.71	3.69
Had at least one MICAP incident		0.39	0.49
Was under a PSCM contract		0.36	0.48
Flying hours (000s)		36.29	28.33
Missing flying hours		0.01	0.12
Engines NIINs			
NIINs	146		
Observations	3,650		
MICAP incidents		2.03	4.63
Had at least one MICAP incident		0.44	0.50
Was under a PSCM contract		0.33	0.47

Figure 4.1
Population Density Function for Actual Data and for an Equivalent Normal Distribution

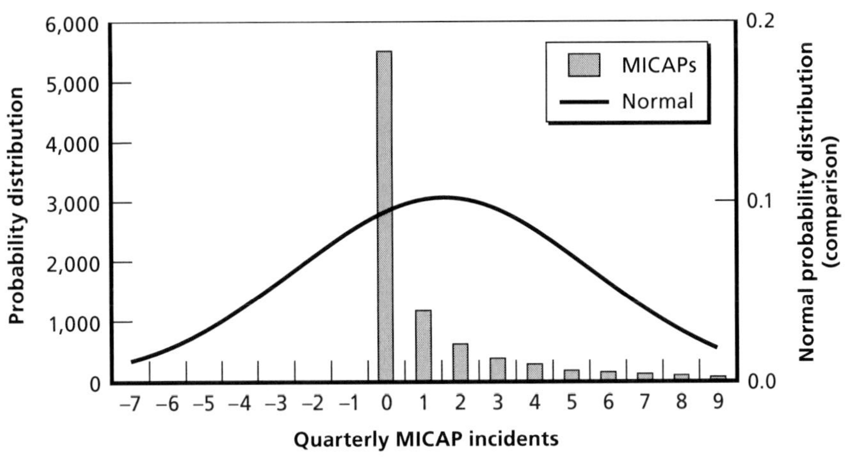

NOTE: The mean is 1.62 and the standard deviation is 3.96.
RAND *MG584-4.1*

clustering occurs at the low MICAP values, especially at zero, with 85 percent of all NIIN-quarter observations having no MICAP incidents. Because these are "count" data (i.e., a count of the number of MICAP incidents) and have this distribution with such large clustering at low values, a typical and appropriate model to use for the outcome variable is a negative binomial econometric model. Thus, we used the negative binomial. Negative binomial models exclude any set of observations that have no variation in the outcome (such as those for a particular NIIN). Thus, they automatically exclude the observations from the 267 PSCM NIINs that have no MICAP incidents. For the outcome of whether there were any MICAP incidents, we used OLS, for which we can use all 624 NIINs, including the 267 NIINs with no variation in the outcome.

CHAPTER FIVE

Results from Econometric Models

For each sample (all NIINs, all NIINs with at least one MICAP incident in the 25 quarters, aircraft NIINs, and engine NIINs), we applied the following three or four regressions for the number of quarterly MICAP incidents:

$$Y_{kt} = \gamma\left(PSCM_{kt}\right) + \varepsilon_{kt} \tag{5.1}$$

$$Y_{kt} = \mu_k + \gamma\left(PSCM_{kt}\right) + \varepsilon_{kt} \tag{5.2}$$

$$Y_{kt} = \mu_k + \mu_t + \gamma\left(PSCM_{kt}\right) + \varepsilon_{kt} \tag{5.3}$$

$$Y_{kt} = \mu_k + \mu_t + \gamma\left(PSCM_{kt}\right) + \lambda\left(C_{kt}\right) + \varepsilon_{kt} \tag{5.4}$$

Equation 5.1 is a simple regression, with PSCM as the only explanatory variable, along with a constant (intercept) term. Equation 5.2 adds NIIN-level fixed effects. Equation 5.3 adds quarter-level fixed effects to the second model. Equation 5.4 adds flying hours (e.g., for a particular NIIN), along with an indicator for missing information on flying hours, both of which are in vector C_{kt}. We could only estimate the fourth model for the aircraft sample because we did not have any co-occurring factors (such as operating hours) for the other types of NIINs. The different regressions demonstrate the importance of including the NIIN-level fixed effects, time-level fixed effects, and co-occurring factors.

Results with Negative Binomial Models

Table 5.1 shows the results from the negative binomial models (Equations 5.1 through 5.4). The model does not incorporate groups of variables for which there is no variation in the outcome. Thus, the model automatically includes only the observations for the 357 NIINs that have at least one MICAP incident in any of the 25 quarters. We excluded the models for the set of all PSCM NIINs because the results are the same as the sample of the NIINs with at least one MICAP incident. For this sample, the coefficient estimates on the post-PSCM variable are all negative and statistically significant. The estimates suggest PSCM contracts are associated with about 0.25 to 0.30 fewer MICAP incidents per quarter (with estimates ranging from –0.252 to –0.305); the mean MICAP incident rate is 1.62. The magnitude of the estimates changes slightly across the specifications but remains statistically significant.

This is not the case for the separate samples for aircraft and engine PSCM NIINs. For engine NIINs, the estimates are also statistically significant, but adding the quarter-level fixed effects has a large influence on the estimated effect of the PSCM initiative on the number of quarterly MICAP incidents, nearly cutting the estimate in half, from –0.475 to –0.230. This difference is statistically significant at the 5-percent level. However, the final estimate does remain statistically significant.

For aircraft NIINs, the estimates also are affected by adding the quarter-level fixed effects, as well as the NIIN-level fixed effects. The estimated effect of the PSCM initiative is not statistically significant until the final model that adds the co-occurring factor of flying hours (Table 5.2). In that case, the estimate is significant at the 10-percent level. The coefficient estimate on flying hours is negative (albeit insignificant), which is contrary to what we expected. The large change in

Table 5.1
Coefficient Estimates for the Negative Binomial Models to Estimate the Number of MICAP Incidents in a Quarter

	Fixed Effects in MICAP Regressions			
	None	**NIIN**	**NIIN and Quarter**	**NIIN, Quarter, and Flying Hours**
Calculated using equation	5.1	5.2	5.3	5.4
All PSCM NIINs (N = 15,600)				
Post-PSCM	N/A	N/A	N/A	N/A
All PSCM NIINs with MICAP incidents[a] (N = 8,925)				
Post-PSCM	−0.299***	−0.305***	−0.252***	
	(0.050)	(0.037)	(0.059)	
Engine NIINs (N = 3,650)				
Post-PSCM	−0.549***	−0.475***	−0.230**	
	(0.050)	(0.054)	(0.093)	
Aircraft NIINs (N = 3,200)				
Post-PSCM	0.017	−0.078	−0.129	−0.163*
	(0.081)	(0.057)	(0.089)	(0.090)
Flying hours (thousands)				−0.003
				(0.002)
Missing flying hours				−1.021***
				(0.002)

NOTES: The dependent variable is the number of quarterly MICAP incidents. Standard errors are in parentheses. Negative binomial models do not produce R^2 statistics.

[a] Includes at least one MICAP incident in at least one of the 25 quarters.

*** Indicates statistical significance at the 1-percent level.

** Indicates statistical significance at the 5-percent level.

* Indicates statistical significance at the 10-percent level.

Table 5.2
Coefficient Estimates from the OLS Models to Estimate the Incidence of Any MICAP Incidents in a Quarter

	Fixed Effects in MICAP Regressions			
	None	**NIIN**	**NIIN and Quarter**	**NIIN, Quarter, and Flying Hours**
Calculated using equation	5.1	5.2	5.3	5.4
All PSCM NIINs (N = 15,600)				
Post-PSCM	−0.056***	−0.036***	−0.042***	
	(0.007)	(0.005)	(0.009)	
Within-NIIN R^2		0.003	0.005	
All PSCM NIINs with MICAP incidents[a] (N = 8,925)				
Post-PSCM	−0.047***	−0.066***	−0.075***	
	(0.011)	(0.010)	(0.016)	
Within-NIIN R^2		0.006	0.009	
Engine NIINs (N = 3,650)				
Post-PSCM	−0.091***	−0.100***	−0.024	
	(0.017)	(0.015)	(0.029)	
Within-NIIN R^2		0.013	0.023	
Aircraft NIINs (N = 3,200)				
Post-PSCM	0.037**	−0.015	−0.064***	−0.063***
	(0.018)	(0.015)	(0.024)	(0.025)
Flying hours (thousands)				0.005***
				(0.002)
Missing flying hours				−0.052
				(0.090)
Within-NIIN R^2		0.001	0.018	0.021

NOTES: The dependent variable is the number of quarterly MICAP incidents. Standard errors are in parentheses. Negative binomial models do not produce R^2 statistics.

[a] Includes at least one MICAP incident in at least one of the 25 quarters.

*** Indicates statistical significance at the 1-percent level.

** Indicates statistical significance at the 5-percent level.

* Indicates statistical significance at the 10-percent level.

the coefficient estimate on the PSCM variable when flying hours are added suggests that flying hours may be correlated with some other co-occurring factors that we could not account for with just the quarter-level fixed effects.

Results with Ordinary Least Squares Models

Table 5.2 shows the results from OLS regressions for which the dependent variable is "whether the NIIN had at least one MICAP incident in a quarter."[1] The coefficient estimates are generally similar in direction to those in Table 5.1 on the number of quarterly MICAP incidents, suggesting that being part of a PSCM contract is associated with a lower probability of having a MICAP incident in a quarter. For the sample of all PSCM NIINs, the estimate on the third regression, which includes both NIIN and quarter-level fixed effects, suggests that being part of a PSCM contract is associated with a 4.2-percentage-point lower probability of having a MICAP incident in a quarter. The estimated effect is a 7.5-percentage-point reduction for the second sample. However, for the engine NIINs, the estimate is no longer statistically significant if the quarter-level fixed effects are included. The estimate for the aircraft NIINs remains significant and is hardly affected by the flying hours. And in this model, the flying hours variable has a significant coefficient estimate, which has the expected positive sign, because we would expect to see more MICAPs as flying hours increase.

The model does not appear to explain much of the variation in MICAP incidents, as the within-NIIN R^2s are very small, which indi-

---

[1] We use OLS models again. Probit or logit models are often used for regressions with a dichotomous outcome. These are nonlinear regressions that are typically used to account for nonlinear effects (especially around very low and very high probabilities) and to avoid predictions outside of the zero-one range. However, we use OLS models, which are linear, rather than probit or logit models, for three reasons: (1) the estimates represent the actual change in the dependent variable from a change in PSCM status from not covered to covered, unlike the probit and logit models; (2) our statistical package (Stata) could not identify the marginal effects with probit and logit models, likely due to the large number of NIIN fixed effects; and (3) we can easily calculate within-NIIN R^2s, which are indicators of the extent of within-NIIN variation that is explained by our explanatory variables.

cates that these regressions are not explaining a large part of the variation in MICAP incidents.[2] For the sample with all NIINs, less than 1 percent of the within-NIIN variation is explained with the PSCM variable. Adding the quarter-level fixed effects does little to improve this, increasing only from 0.003 to 0.005. The within-NIIN R^2 for the aircraft NIINs explains even less than that for the full set of NIINs (about 0.1 percent) when not including quarter-level fixed effects. It does increase much more than the full sample of NIINs when quarter-level fixed effects are included (0.018). Further, when we added the co-occurring factor (flying hours) to the model, the R^2 increased only to 0.021. Thus, this co-occurring factor explains only a small portion of the variation in the outcome. This indicates, at least for an analysis of MICAP incidents, that better co-occurring factors need to be identified.

Other Models Explored

The model we developed for MICAP incidents has other applications. We also explored estimating similar relationships between PSCM and awaiting parts (AWP) incidents and between PSCM and contract prices.[3] We found only 46 Air Force–managed PSCM NIINs that had AWP incidents over three fiscal years of AWP data (FYs 2002–2004), so this would not have been very fruitful, especially relative to the MICAP models. The Defense Logistics Agency manages many AWP items, but because of time and resource constraints, we had to limit our spending data to Air Force purchases. If we had included Defense Logistics Agency purchases, the available sample of AWP observations would have increased considerably.

With contract data, we also attempted to estimate how the same PSCM initiatives affected contract prices for specific NIINs. We were

[2] The R^2 indicates the proportion of the variation in the dependent variable that is explained by the explanatory variables.

[3] In an *AWP incident*, repair of a larger component part is delayed because one or more parts necessary for its repair are unavailable.

unable to obtain stable estimates in the model, as successively excluding large outliers from the sample caused very large changes in the estimated effects of the PSCM initiative. In addition, poor data quality may be causing problems; we observed evidence of significant amounts of potentially miscoded data—many observations appeared to differ by factors of 100 or 0.01. For example, one specific NIIN cost $5 one year and $500 the next year. Unfortunately, determining the accuracy of the observations was beyond the scope of this analysis.

As more data become available in the future, this approach may be useful for more applications, depending on the quality of the data. Data of questionable quality, however, could make such applications difficult to implement.

CHAPTER SIX

Conclusions

This monograph has described an effort to help the Air Force determine whether and how much its PSCM initiatives are actually improving supply-chain outcomes. But making such determinations when many related changes are happening at the same time makes properly attributing causality a major challenge. The models we have described here explore ways of addressing these "co-occurring factors."

That effort broke down into three parts:

1. determining what factors exist that change over time, are concurrent with the PSCM initiatives, and might affect PSCM metrics
2. obtaining suitable data for these factors
3. controlling for them.

Curiously, we found that literature on tracking supply-chain metrics and on estimating the outcomes of supply-chain initiatives did not adequately address this problem. We therefore chose to develop our own generic regression-econometric model for NIIN-level metrics that could theoretically control for co-occurring factors, provided that there are adequate and accessible data.

In an initial test of the method developed, we applied several versions of the regression model to an analysis of quarterly MICAP incidents. The results demonstrated the importance of including both NIIN- and time-level fixed effects. The time-level fixed effects can be considered co-occurring factors that are general to the set of NIINs being examined, so the fact that including these effects influenced the estimates on the PSCM indicator demonstrates how unobserved co-occurring factors can confound the estimated outcomes of PSCM initiatives. Furthermore, even though we had a co-occurring factor (flying hours) that is weakly associated with NIIN removals, we still found that it, too, influenced the estimated effect of the PSCM initiative. Because including time-level fixed effects and flying hours influenced the estimated PSCM effect, other unknown co-occurring factors specific to individual NIINs are likely confounding the estimated PSCM initiative outcomes. Thus, conclusions drawn from our finding that the PSCM contracts reduced the number of quarterly MICAP incidents should take into account the possibility that the estimated effects captured the effects of co-occurring factors for which we had not adequately accounted.

The model we analyzed had very low explanatory power (R^2)—that is, it explained a small portion of the variation in the dependent variable, which is a test of how well the model explains a certain kind of observable outcome. The low explanatory power could have several explanations. First, we likely have a weak or inadequate co-occurring factor (flying hours). Second, PSCM is one of many factors that affect the performance of the supply chain. Unfortunately, many of these are internal to the supply chain (such as the number of repairs in a quarter), so incorporating them into an analysis of MICAP incidents would require more-advanced econometric models, which are highly dependent on certain assumptions about what variables affect what metrics. Third, the simple dichotomous indicator for PSCM may be too general. Quantifying more detail on what was implemented on each PSCM might make it possible to learn more about the effectiveness of particular PSCM practices.

We believe that the major hurdle that the Air Force faces in estimating the true effectiveness of PSCM initiatives using various metrics is having adequate data on co-occurring factors. In our analysis of MICAP incidents, an ideal co-occurring factor would be the number of NIIN removals, because it is the primary precursor to a MICAP incident and is external to the supply chain (exogenous). Any MICAP incidents resulting from part removals could be due, for instance, to other unobserved supply-chain effects, such as changes in inventory policies or maintenance capacities. The best co-occurring factor for which we were able to obtain data was the number of flying hours for the end item most associated with the particular NIIN. While the estimates for the most part changed with inclusion of the flying hours, the link is still weak between flying hours and MICAP incidents (based on the R^2). Therefore, flying hours is not an adequate co-occurring factor, which means that our estimates of the benefits of the PSCM initiative may be incorrectly estimated. Future analyses can also use the number of sorties associated with a NIIN as a co-occurring factor; these may be more important for some NIINs than the number of flying hours (Slay and Sherbrooke, 1997; Rainey, 2001). Analysts could also consult product select codes in determining the best co-occurring factors for particular NIINs.

We believe that good data on co-occurring factors (such as NIIN removals for an analysis of MICAP incidents) would be necessary to identify the true effects of these PSCM initiatives.

The methodology developed in this monograph can be applied to the evaluation of other programs and initiatives. However, to obtain accurate estimates of the benefits of these initiatives, care must be taken to ensure that relevant co-occurring factors are accounted for. Doing so would permit a test of the usefulness of this approach to estimating the effect of PSCM or other initiatives on metrics that are aligned with the goals of other initiatives, such as eLog21.

APPENDIX

PSCM Contract Data

Table A.1
Information on the 17 PSCM Contracts Used for Contract Data

Company	Contract	Contract Value ($000s)	NIINs (no.)	Date of First Contract	Two-Position Federal Supply Class
Pratt & Whitney	SP040001D9405	518,372	158	Sept. 2002	Engines, Turbines, and Comp.
GE	F4160800D0323	189,407	94	Oct. 2000	Engines, Turbines, and Comp.
Boeing	F0960301D0025	137,780	55	Apr. 2001	AC and AF Struct. Comp.
Honeywell	F3460100D0371	56,302	90	Jan. 2001	AC Comp. and Accessories
Raytheon (Goleta)	FA852304D0001	52,877	5	Feb. 2004	Elect. and Elect. Equip. Comp.
Rolls Royce	F3460101D0155	51,982	25	Jul. 2001	Engines, Turbines, and Comp.
Parker Hannifin	F3460101D0228	45,291	47	Sept. 2001	Valves
Goodrich	F4263002D0011	30,349	56	Aug. 2002	AC Comp. and Accessories
BAE Systems	F0960303D0001	16,884	25	May 2003	Comm., Detect. and Coherent Rad. Equip.
General Dynamics	F4262000D0097	11,794	50	Sept. 2000	Elect. and Elect. Equip. Comp.

Table A.1—Continued

Company	Contract	Contract Value ($000s)	NIINs (no.)	Date of First Contract	Two-Position Federal Supply Class
Aircraft Braking Systems	SP040002D9403	6,302	9	Apr. 2003	AC Comp. and Accessories
Raytheon	F0960302D0101	5,938	18	Jun. 2003	Elect. and Elect. Equip. Comp.
Northrup	F0960303D0002	5,706	15	Sept. 2003	Elect. and Elect. Equip. Comp.
Nassau Tool Works	FA820304D0004	874	1	Feb. 2004	AC Comp. and Accessories
Castle Precision	FA820304D0008	325	2	Feb. 2004	AC Comp. and Accessories
CCC/Heroux-Devtek	FA820304D0012	33	1	May 2004	AC Comp. and Accessories
Rockwell Collins	FA852304D0002	27	3	Aug. 2004	Elect. and Elect. Equip. Comp.

SOURCE: Global Combat Suport System—Air Force (formerly Air Force Knowledge Service), Strategic Sourcing Analysis Tool, FYs 2001 to 2004. These 17 contracts include two PSCM contracts and 15 "pre–commodity council," that is, corporate and strategic-sourcing contracts.

Bibliography

Adams, John L., John Abell, and Karen E. Isaacson, *Modeling and Forecasting the Demand for Aircraft Recoverable Spare Parts*, Santa Monica, Calif.: RAND Corporation, R-4211-AF/OSD, 1993. As of June 11, 2007: http://www.rand.org/pubs/reports/R4211/

AFMC—*See* U.S. Air Force, Air Force Materiel Command.

Arkes, Jeremy, "Does the Economy Affect Teenage Substance Use?" *Health Economics*, Vol. 16, No. 1, January 2007, pp. 19–36.

Arkes, Jeremy, and M. Rebecca Kilburn, *Modeling Reserve Recruiting: Estimates of Enlistments*, Santa Monica, Calif.: RAND Corporation, MG-202-OSD, 2005. As of August 7, 2007: http://www.rand.org/pubs/monographs/MG202/

Ausink, John, Laura H. Baldwin, Sarah Hunter, and Chad Shirley, *Implementing Performance-Based Services Acquisition (PBSA): Perspectives from an Air Logistics Center and a Product Center*, Santa Monica, Calif.: RAND Corporation, DB-388-AF, 2002. As of June 22, 2007: http://www.rand.org/pubs/documented_briefings/DB388/

Ausink, John, Frank Camm, and Charles Cannon, *Performance-Based Contracting in the Air Force: A Report on Experiences in the Field*, Santa Monica, Calif.: RAND Corporation, DB-342-AF, 2001. As of June 22, 2007: http://www.rand.org/pubs/documented_briefings/DB342/

Avery, Susan, "Rockwell Collins Takes Off," *Purchasing*, Vol. 132, No. 3, February 20, 2003, pp. 25–28.

———, "Lean, but Not Mean, Rockwell Collins EXCELS," *Purchasing*, Vol. 134, No. 14, September 1, 2005, pp. 26–32.

Baldwin, Laura H., Frank Camm, and Nancy Y. Moore, *Strategic Sourcing: Measuring and Managing Performance*, Santa Monica, Calif.: RAND Corporation, DB-287-AF, 2000. As of June 22, 2007: http://www.rand.org/pubs/documented_briefings/DB287/

———, *Federal Contract Bundling: A Framework for Making and Justifying Decisions for Purchased Services*, Santa Monica, Calif.: RAND Corporation, MR-1224-AF, 2001. As of June 22, 2007:
http://www.rand.org/pubs/monograph_reports/MR1224/

Chenoweth, Mary E., and Clifford Grammich, *F100 Engine Purchasing and Supply Chain Management Demonstration: Findings from Air Force Spend Analyses*, Santa Monica, Calif.: RAND Corporation, MG-424-AF, 2006. As of June 22, 2007:
http://www.rand.org/pubs/monographs/MG424/

Dixon, Lloyd, Chad Shirley, Laura H. Baldwin, John A. Ausink, and Nancy F. Campbell, *An Assessment of Air Force Data on Contract Expenditures*, Santa Monica, Calif.: RAND Corporation, MG-274-AF, 2005. As of June 22, 2007:
http://www.rand.org/pubs/monographs/MG274/

Dryden, Sue, and Marie Tinka, "PSCM/DMT [Depot Maintenance Transformation] Update," presentation for AFMC at the Sustainment Transformation Senior Leadership Conference III, Atlanta, Ga., September 9, 2004.

Gabreski, Lt Gen Terry L., Supply Chain Council Award for Excellence, Oklahoma City Air Logistics Center, Air Force Materiel Command, Tinker Air Force Base, Oklahoma City, OK, 2004. As of November 23, 2006:
http://www.acq.osd.mil/log/sci/awards/2004_award/AFMC_OK_ALC_Expeditionary_Logistics_for_the_21st_century_a.pdf

Gorski, Mathew J., *The Application of an Army Prospective Payment Model Structured on the Standards Set Forth by the CHAMPUS Maximum Allowable Charges and the Center for Medicare and Medicaid Services: An Academic Approach*, Ft. Sam Houston, Tex.: Brooke Army Medical Center, HCA-34-05, April 29, 2005.

Greene, William H., *Econometric Analysis*, 5th ed., New York: Prentice Hall, 2002.

Gunasekaran, A., C. Patel, and E. Tirtiroglu, "Performance Measures and Metrics in a Supply Chain Environment," *International Journal of Operations & Production Management*, Vol. 21, No. 1/2, 2001, p. 71.

Hillestad, Richard J., *Dyna-METRIC: Dynamic Multi-Echelon Technique for Recoverable Item Control*, Santa Monica, Calif.: RAND Corporation, R-2785-AF, 1982. As of June 11, 2007:
http://www.rand.org/pubs/reports/R2785/

Johnson, Clay III, Deputy Director of Management, "Implementing Strategic Sourcing," memorandum, Washington D.C.: Office of Management and Budget, May 20, 2005. As of June 2, 2006:
http://www.whitehouse.gov/omb/procurement/comp_src/implementing_strategic_sourcing.pdf

Kaplan, Robert S., and David P. Norton, "The Balanced Scorecard: Measures That Drive Performance," *Harvard Business Review*, Vol. 70, No. 1, 1992, pp. 71–78.

Kem, Dale A., Bobby Jackson, Avery Williams, and Virginia Stouffer, *Performance Metrics for Defense Working Capital Funds: A Focus on Supply Management*, McLean, Va.: Logistics Management Institute, LMI-PA804T1, July 2000.

Klapper, Larry S., Neil Hamblin, Linda Hutchison, Linda Novak, and Jonathan Nivar, *Supply Chain Management: A Recommended Performance Measurement Scorecard*, McLean, Va.: Logistics Management Institute, LG803R1, June 1999.

Koenig, E., *AFMC Supply Chain Metrics Guide*, Langley, Va.: Logistics Management Institute, 80067836-1027031, 2003.

Leatham, Mark, "eLog21—Purchasing and Supply Chain Management," *Air Force Journal of Logistics*, Winter 2003.

Moore, Nancy Y., Cynthia Cook, Clifford Grammich, and Charles Lindenblatt, *Using a Spend Analysis to Help Identify Prospective Air Force Purchasing and Supply Management Initiatives: Summary of Selected Findings*, Santa Monica, Calif.: RAND Corporation, DB-434-AF, 2004. As of June 22, 2007: http://www.rand.org/pubs/documented_briefings/DB434/

Moore, Nancy Y., Laura H. Baldwin, Frank Camm, and Cynthia R. Cook, *Implementing Best Purchasing and Supply Management Practices: Lessons from Innovative Commercial Firms*, Santa Monica, Calif.: RAND Corporation, DB-334-AF, 2002. As of June 11, 2004: http://www.rand.org/pubs/documented_briefings/DB334/

Nicosia, Nancy, and Nancy Moore, *Implementing Purchasing and Supply Chain Management: Best Practices in Market Research*, Santa Monica, Calif.: RAND Corporation, MG-473-AF, 2006. As of June 22, 2007: http://www.rand.org/pubs/monographs/MG473/

Rainey, James C., Robert McGonagle, Beth F. Scott, and Gail Waller, *USAF Maintenance Metrics Handbook*, Maxwell Air Force Base, Ala.: Air Force Logistics Management Agency, December 20, 2001. As of December 5, 2006: http://www.aflma.hq.af.mil/lgj/Maintenance%20Metrics%20Handbook.pdf

Reese, David, and Mark Hansen, "Commodity Council Concept of Operations," Washington, D.C.: Headquarters, U.S. Air Force, January 2003.

Roche, James E., and John P. Jumper, *Expeditionary Logistics for the 21st Century Campaign Plan*, Headquarters United States Air Force, April 1, 2005.

Rockwell Collins, Inc., Form 10-K for the Fiscal Year Ended September 30, 2002. As of May 2006: http://yahoo.brand.edgar-online.com/fetchFilingFrameset.aspx?FilingID=4032133&Type=HTML

———, Form 10-K for the Fiscal Year Ended September 30, 2005. As of May 2006:
http://yahoo.brand.edgar-online.com/fetchFilingFrameset.aspx?FilingID=2091078&Type=HTML

Ruhm, C., "Economic Conditions and Alcohol Problems," *Journal of Health Economics*, Vol. 14, 1995, pp. 583–603.

Slay, F. Michael, and Craig C. Sherbrooke, *Predicting Wartime Demand for Aircraft Spares*, McLean, Va.: Logistics Management Institute, AF501MR2, April 1997.

Stewart, Gordon, "Supply Chain Performance Benchmarking Study Reveals Keys to Supply Chain Excellence," *Logistics Information Management*, Vol. 8, No. 2, 1995, pp. 38–44.

Streeter, Melanie, "Air Force Logistics Moves into New Century with 'eLog21,'" *Air Force Print News*, December 24, 2003.

U.S. Air Force, *Improving Air and Space Equipment Reliability and Maintainability, Maintenance*, Air Force Instruction 21-118, 2 October 2003. As of November 21, 2006:
http://www.e-publishing.af.mil/pubfiles/af/21/afi21-118/afi21-118.pdf

———, *Expeditionary Logistics for the 21st Century Campaign Plan*, Washington, D.C., 2005. As of November 13, 2006:
http://www.af.mil/shared/media/document/AFD-060831-041.pdf

———, "Commodity Council Implementation and Operations," Air Force Federal Acquisition Regulation Supplement, IG5307.104-93, June 2006. As of December 5, 2006:
http://farsite.hill.af.mil/reghtml/regs/far2afmcfars/af_afmc/affars/IG5307.104-93.htm

———, *Basic USAF Supply Manual*, Washington, D.C., Air Force Manual 23 110, April 1, 2007. As of June 26, 2007:
http://www.e-publishing.af.mil/pubfiles/af/23/afman23-110/afman23-110.pdf

U.S. Air Force, Air Force Materiel Command, "Improving Warfighter Readiness Through PSCM Transformation," PSCM fact sheet, October 6, 2004. As of June 12, 2007:
http://www.af.mil/shared/media/document/AFD-060831-043.pdf

———, "Requirements for Secondary Items (D200A, D200N)," AFMC Manual 23-1, March 16, 2005. As of December 4, 2006:
http://www.e-publishing.af.mil/pubfiles/afmc/23/afmcman23-1/afmcman23-1.pdf

———, "CRM Concept of Operations," Version 1.1, Wright Patterson Air Force Base, February 9, 2006. As of December 4, 2006:
https://wwwd.my.af.mil/afknprod/Database/Staging/2650056/CRM_CONOPS_Vfin.doc

U.S. General Accounting Office, *Air Force Depot Maintenance: Management Changes Would Improve Implementation of Reform Initiatives*, Washington, D.C., NSIAD-99-63, 1999.[1]

---

[1] The GAO is now known as the Government Accountability Office.